CAIRNGORMS

A LANDSCAPE FASHIONED BY GEOLOGY

SCOTTISH
NATURAL
HERITAGE

©Scottish Natural Heritage 2006
Second edition
ISBN 1 85397 455 2
A CIP record is held at the British Library
IA3K0406

Acknowledgements
Authors: John Gordon, Rachel Wignall, Ness Brazier and Patricia Bruneau
Series editor: Alan McKirdy (SNH)

Photography:
David Bell/ECOS 25; **Steve Austin** 37 left; **British Geological Survey** 5, 9 right, 12;
BGS enhanced image © NERC 2005 4; **British Antarctic Survey** 16; **Basil Dunlop** 35 right;
Lorne Gill/SNH front cover, back cover, frontspiece, 6, 8, 10 left, 10 right, 19, 21 top, 21 bottom, 26, 27
right, 28, 30, 31 top, 31 bottom, 33, 34, 35 left; **John Gordon** 15, 17, 23 top, 24, 29, 36; **Patricia &
Angus Macdonald/SNH** 13, 18, 20, 27 left; **David Whitaker** 37 right; **Rachel Wignall** 14, 23 bottom

Illustrations:
Craig Ellery 2, 3, 6, 7, 9, 22, 32, 33; **Clare Hewitt** 11; **Jim Lewis** 17; **Iain McIntosh** contents

Further copies of this book and other publications can be obtained from:

The Publications Section,
Scottish Natural Heritage,
Battleby, Redgorton, Perth PH1 3EW
Tel: 01738 444177 Fax: 01738 827411
E-mail: pubs@snh.gov.uk
www.snh.org.uk

Front cover image:
Glen Quoich
Back cover image:
Tree stumps preserved in peat bogs record
the former extent of native woodland

CAIRNGORMS

A Landscape Fashioned by Geology

by

John Gordon, Rachel Wignall, Ness Brazier and Patricia Bruneau

Panorama of Glenmore and the northern Cairngorms

Contents

*N*ot less impressive than the corries, the cliffs, and the dark lakes are the great wastes of shattered stone and sand which form the summit plateaus and ridges of these mountains ... *The very bareness of these mountain-tops is on a majestic scale, and it forms one of the elements in the massive grandeur and repose which are the distinguishing characteristic of the Cairngorms.*
Sir Henry Alexander. 1928. *The Cairngorms.* Scottish Mountaineering Club.

The high granite plateaux, steep-sided glens and deep corries make the Cairngorms one of the most renowned and distinctive mountain landscapes in Britain. They form the largest area of high ground in the country, and with their severe climate, the mountains have a strong 'arctic' character. The rocks, landforms and soils of the Cairngorms, along with the climate, have exerted a powerful influence on the landscape, wildlife and land use of the area.

Thousands of people visit the Cairngorms each year to walk, climb and ski, and to appreciate the magnificent scenery and wildlife. Few, perhaps, are aware of the geological and geomorphological processes which have shaped this unique environment and which underlie its exceptional importance for nature conservation. In this booklet we trace the evolution of the Cairngorms and discover how the rocks, landforms and soils have been fashioned through time by weathering, glaciers and rivers, and how the natural environment, climate and plant life have changed.

The Cairngorms Through Time

QUATERNARY (THE ICE AGE) 2.6 million years ago to present		**11,500 years ago to the present day.** The climate warmed rapidly at start of the present interglacial (the Holocene). Soils stabilised and pioneer vegetation was followed later by the spread of pine forest about 9,000 years ago. River terraces formed and river channels changed positions during floods. Human activity resulted in significant clearance of the natural woodland during the last 6,000 years. Today, periglacial processes continue to modify the higher slopes and summits, and debris flows and floods periodically alter the slopes and floors of the glens. **12,900 to 11,500 years ago.** The climate became intensely cold again at the time of the Loch Lomond Readvance. Small mountain glaciers re-formed and produced moraines in some corries. Slopes were extensively modified by solifluction and rockfalls. **15,000 to 12,900 years ago.** The climate warmed rapidly, with summer temperatures similar to those of today; most if not all of the remaining glaciers melted. Pioneer plants colonised the glacial deposits and the soil stabilised; arctic shrub tundra developed, with dwarf birch and crowberry. **18,000 to 15,000 years ago.** Retreat and local readvances of the last ice sheet resulted in the formation of moraines, meltwater channels, kame terraces, eskers, kettle holes, ice-dammed lakes and deltas. **30,000 to 18,000 years ago.** The last (Late Devensian) ice sheet expanded, reaching its maximum extent about 22,000 years ago or slightly earlier; the glaciers further enlarged the major features of glacial erosion. **2.6 million years to 30,000 years ago.** The onset of the Ice Age occurred about 2.6 million years ago. There were many glacial (cold) and interglacial (warmer) episodes with repeated growth and decay of glaciers - mainly mountain (including corrie) glaciers before about 750,000 years ago and thereafter a succession of large ice sheets. Major features of glacial erosion were formed, including glacial troughs, breached watersheds and corries. Further, shallower, weathering of the bedrock took place, with stripping of the weathered rock and exhumation of the present tors.
NEOGENE 23 to 2.6 million years ago		Weathering continued under warm, temperate conditions. Further uplift of the area and erosion of the glens occurred. Cooling of the climate intensified after about 3 million years ago.
PALAEOGENE 65 to 23 million years ago		Chemical weathering of the granite took place under humid, sub-tropical conditions. Periods of uplift and erosion of the glens and straths occurred, and the plateau surfaces were formed. The Cairngorms were re-established as an upland area. The North Atlantic Ocean opened to the west of Scotland.
CRETACEOUS 146 to 65 million years ago		Warm, shallow seas covered most of Scotland. The Cairngorms were probably a low area rising above sea level.
JURASSIC 200 to 145 million years ago		The Cairngorms remained an upland area, with rivers draining to deltas in the North Sea basin.
TRIASSIC 251 to 200 million years ago		The Cairngorms remained an upland area, but with reduced relief, surrounded by arid continental plains. Scotland lay 20 to 30 degrees north of the Equator at this time.
PERMIAN 299 to 251 million years ago		The Cairngorms remained an upland area and desert sands accumulated along the present-day margins of Scotland.
CARBONIFEROUS 359 to 299 million years ago		Scotland drifted north into equatorial latitudes. Deposits of coal formed in low-lying coastal swamps in the Midland Valley, but the Cairngorms remained an upland area. Rivers continued to erode the main lines of weakness in the granite.
DEVONIAN 416 to 359 million		Intensive erosion of the Caledonian mountains produced vast thicknesses of sand and gravel which were deposited by rivers in semi-arid basins within the mountains and in adjacent areas of the Midland Valley. After the Cairngorm granite was exposed at the ground surface, rivers selectively eroded the main lines of weakness in the granite, forming the precursors of the present glens. Scotland at this time lay about 10 degrees south of the Equator.
SILURIAN 444 to 416 million years ago		The continents of Baltica (Scandinavian Europe) and Eastern Avalonia (containing England) collided with Laurentia as the Iapetus Ocean closed completely. The Cairngorm granite rose as a molten mass from deep in the Earth's crust to within several kilometres of the surface. By 427 million years ago, the Cairngorm granite had cooled and solidified.
ORDOVICIAN 488 to 444 million years ago		Formation of the Caledonian mountains as the Iapetus Ocean closed and a chain of volcanic islands collided with Laurentia, the continent on which Scotland was then located. The collision was accompanied by metamorphism and folding of the Dalradian sediments.
CAMBRIAN 542 to 488 million years ago		Dalradian sediments continued to be deposited in the Early Cambrian. In Late Cambrian times, the Iapetus Ocean began to close.
PRECAMBRIAN Before 542 million years ago		**After 800 million years ago.** Dalradian sediments began to be deposited in shallow seas and later on the edge of the widening Iapetus Ocean. **4000 million years ago.** Formation of the oldest known rocks on Earth. **4500 million years ago.** Formation of the Earth.

Geological Map of the Cairngorms

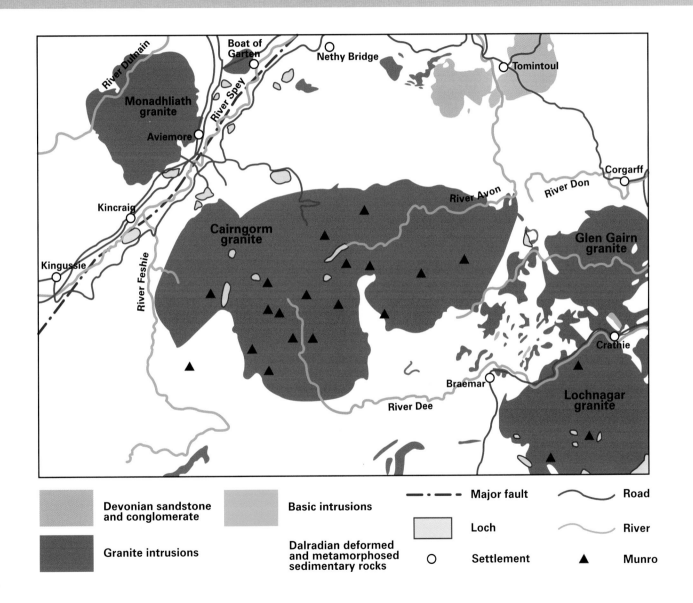

Devonian sandstone and conglomerate

Basic intrusions

Granite intrusions

Dalradian deformed and metamorphosed sedimentary rocks

–·–·– Major fault

Loch

○ Settlement

〰 Road

〰 River

▲ Munro

A Changing Landscape

Satellite image of the Cairngorms

The landscape of the Cairngorms has a remarkable history stretching far back in geological time. The processes that have fashioned this old landscape can be traced today in the rocks, landforms and soils beneath our feet and in the shapes of the mountains.

The story of the Cairngorms begins over 700 million years ago, when muds and sands were deposited in a vast ocean.

Some 250 million years later, the ocean disappeared as ancient continents collided during a time of great geological upheaval caused by the movements of the plates that make up the Earth's surface. The collisions also resulted in the uplift of a mountain range of Himalayan proportions and the emplacement of a mass of granite deep in the Earth's crust. Over many millions of years, weathering and erosion levelled the ancient mountains and laid bare their granite roots.

Plateau surface of the high Cairngorms looking west from the summit of Cairn Gorm

Once at the surface, the granite rocks that are now the trademark of the Cairngorms were in turn sculpted by natural processes to form the present mountain range. These processes were accompanied by dramatic shifts in climate. For a time around 50 million years ago, the granite was deeply weathered when subtropical conditions existed in Scotland. Then more recently, during the Ice Age of the last 2.6 million years, glaciers repeatedly formed in the mountains and added their own distinctive signature to the landscape. This is clearly seen today in the shapes of the corries and glens and in the patterns of lochs and low hills in the glens and straths. Since the glaciers melted, rivers and slope processes have continued to modify the landscape. During the last few thousand years, human activities have increasingly become a major force of change, particularly through woodland clearance, accelerated soil erosion and alterations to the rivers. Although the rate of change has decreased since the end of the last glaciation, floods on the Rivers Feshie and Spey nevertheless remind us of the variability and unpredictability of natural processes.

From the record of the rocks and landforms, even though it is incomplete, we can see that the shaping of the Cairngorms has been both long and varied. The overwhelming image is of a dynamic landscape, with dramatic changes in climate and surface processes which have produced a remarkable geodiversity - the variety of rocks, landforms and soils - in a relatively small area. We now turn to reveal in more detail the story of Cairngorms and the journey through time of this exceptional landscape.

Oceans and Colliding Continents

The oldest rocks in the Cairngorms are around 700 million years old, and are known as Dalradian rocks. These rocks underlie much of Scotland between the Great Glen Fault, which defines the line of the Great Glen, and the Highland Boundary Fault, which runs from Stonehaven, through Loch Lomond, to Arran.

The Dalradian rocks formed as immense Earth forces broke apart an ancient continent to create an ocean called the 'Iapetus Ocean'. Present-day Scotland, North America and Greenland were part of a continent known as 'Laurentia' which lay on one side of this ocean. Sediment was deposited at the edge of the ocean in water which periodically deepened due to movement of geological faults. The Dalradian rocks formed from this sediment and include sandstones and quartzite from shallow water and coastal sands; limestones from lime chemically precipitated in coastal lagoons; and siltstones, mudstones, slate, and shales from deeper-water silts, muds and clays.

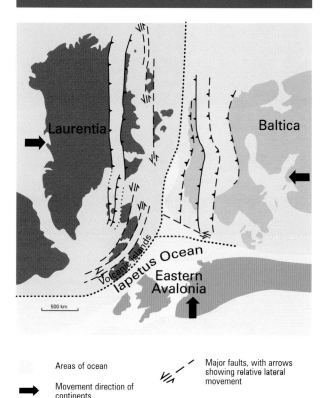

Closure of the Iapetus Ocean and collision of Laurentia, Baltica and Eastern Avalonia

Laurentia

Baltica

Volcanic Islands

Iapetus Ocean

Eastern Avalonia

500 km

Areas of ocean

Movement direction of continents

Collision zones

Major faults, with arrows showing relative lateral movement

'Thrust' faults with barbs on upper, overriding plate

Dalradian rocks exposed in the dramatic gorge at Linn of Dee

Around 200 million years after the Iapetus Ocean began to form, the drifting of the continents started to close it again. For around 100 million years the ocean grew narrower until, around 425 million years ago, it closed completely. As the ocean closed, first a string of volcanic islands, and then two continents, Baltica (Scandinavian Europe) and Eastern Avalonia (containing England), collided with Laurentia.

The first of these continental collisions resulted in the Dalradian sediments being squeezed, crumpled and deformed. The sediments were buried deeply, compressed and heated, turning them into 'metamorphosed' rocks. They were then forced upwards into a huge mountain chain – possibly as big as the present-day Himalayas. The remains of this mountain belt include the Appalachian Mountains in North America and the mountains of Scandinavia, as well as the Grampian Highlands of Scotland.

In the Cairngorms, the majority of the Dalradian rocks are metamorphosed sandstones, including white 'quartzites' from very quartz-rich sands. However, there are also areas with slate and limestone, most notably a broad band to the east of the Cairngorm Mountains passing through Tomintoul and Braemar. The Dalradian rocks are best seen where rivers have cut down and exposed the bedrock. Sometimes this occurs in dramatic gorges such as The Colonel's Bed in Glen Ey and at Linn of Dee.

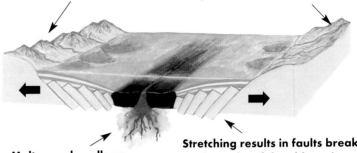

AROUND 600 MILLION YEARS AGO
The Iapetus Ocean was opening and 'Dalradian' sediments were being laid down

Rivers erode the land and deposit sediment in the ocean

Molten rock wells up to form new ocean floor

Stretching results in faults breaking up the rocks of the old continent

AROUND 500 MILLION YEARS AGO
The volcanic islands were on a collision course with Laurentia as the Iapetus Ocean was closing

Volcanic islands

Laurentia

Ocean crust is 'subducted' as the ocean closes

Water in subducted crust causes melting which produces volcanoes

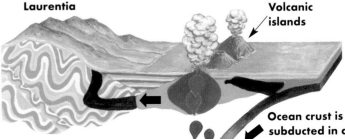

AROUND 460 MILLION YEARS AGO
Collision of the volcanic islands and Laurentia formed a huge mountain chain

Laurentia

Volcanic islands

Deformed and metamorphosed Dalradian sediments form mountains

Ocean crust is subducted in a new location as the Iapetus Ocean continues to close

The Cairngorm Granite

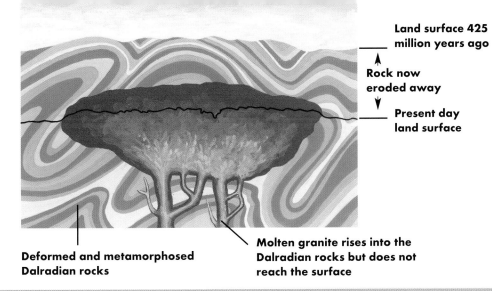

Land surface 425 million years ago

Rock now eroded away

Present day land surface

Deformed and metamorphosed Dalradian rocks

Molten granite rises into the Dalradian rocks but does not reach the surface

Schematic cross section through the Cairngorm granite pluton showing how it formed underground

Hand specimen of Cairngorm granite showing its distinctive pink colour

The Cairngorm granite underlies all of the high land of the Cairngorm Mountains, except for the Mòine Mhór area in the south-west, and is Britain's second largest single area of granite. It formed around 425 million years ago as a result of the continental collisions which deformed the Dalradian rocks after the Iapetus Ocean closed. As well as metamorphosing the Dalradian rock, these collisions caused rock deep down in the Earth's crust to melt. This molten rock, or magma, was lighter than the surrounding rock, so gradually rose upwards, pushing its way into the deformed Dalradian sediments. Much of the magma never reached the surface. Instead, it cooled and solidified several kilometres underground, forming 'igneous intrusions' or 'plutons'.

The majority of the igneous rock in the Cairngorms area is silica-rich granite, intruded after folding of the surrounding Dalradian rocks had ended, although there are some small earlier intrusions of a composition known as 'basic'. The Cairngorm granite is the largest of a cluster of granite plutons which lie along the line of an ancient fault followed by the Dee valley. This fault is thought to be associated with the opening of the Iapetus Ocean, and provided a zone of weakness exploited by the rising magma.

The Cairngorm granite is mostly pink to red in colour but can vary in texture. The variation in texture suggests that it may have formed from a number of distinct 'pulses' of molten rock, each rising to the surface separately, but in relatively quick succession. Each pulse of magma would have had a slightly different composition, or would have cooled at a slightly different rate, giving rise to the variations in appearance.

Although the Cairngorm granite is now exposed at ground level, and some has clearly been removed by erosion, it is believed that the amount of granite removed in this way is relatively small, and that the present-day land surface is not very far below the original 'roof' of the intrusion.

Ancient Rivers and Eroding Mountains

After the end of the continental collisions 425 million years ago, weathering and erosion wore away at the landscape for millions of years. During the Devonian period (416 - 359 million years ago), Scotland lay just south of the Equator and had a hot, semi-arid climate. Material eroded from high, arid mountains was carried away by large rivers and much of it was dumped in low-lying areas and lakes in the mountains and beyond. The rocks formed from these river and lake sediments include conglomerates from pebbles, cobbles and boulders; sandstones from sand; and shales and silts from mud. These Devonian rocks are often called 'Old Red Sandstone' sediments because they are commonly a reddish colour.

Very little Devonian sediment is found in the Cairngorms area. The area of what is now the central massif is thought to have been eroding, and much of the sediment deposited around its margins has now itself been eroded away. However, there are sedimentary rocks of early Devonian age around Tomintoul, where they fill a depression in the underlying Dalradian rocks. The Devonian rocks include red sandstone to the north-east of Tomintoul, and an underlying and much larger area of thick conglomerate. This conglomerate can be seen clearly in the Ailnack Gorge south of Tomintoul, lying directly on top of the much older Dalradian rocks which formed the landscape when the Devonian sediments were deposited.

Devonian conglomerate in a road cutting near the Ailnack Gorge

While the beds of the present-day Water of Ailnack and River Avon contain many granite pebbles and boulders, the pebbles and boulders within the Devonian conglomerate do not include any granite. This tells us that in early Devonian times, when this conglomerate was being formed, the Cairngorm granite was still buried and not yet exposed to erosion. Eventually, however, exposure of the Cairngorm granite did occur, and weathering and erosion started to shape the present granite landscape.

A view of how the landscape around Tomintoul may have looked during the Devonian time period

Geology Shapes the Landscape

Vertical and horizontal joints in granite at the cliff edge of Cairn Lochan

Much of the shape of the landscape in and around the Cairngorms depends on the underlying geology. Greater erosion of weaker rock has given rise to lower-lying areas, while more resistant rock forms the high land and mountains. From Kingussie to Grantown-on-Spey, the wide strath of the River Spey follows a line of weakened rock around a geological fault which runs north-east from Loch Ericht.

The great bulk of the Cairngorms massif is also the product of variations in rock resistance to erosion. The granite is generally more resistant than the surrounding Dalradian rocks and forms the high ground. Some Dalradian rock, however, is also resistant to erosion and forms the high area of the Mòine Mhór.

The steep glens which cut through the granite of the Cairngorms also follow lines of weakness, but they are not faults. They are believed to be associated with alteration of the rock by circulating hot water, a process known as 'hydrothermal alteration', which occurred as the granite cooled. Water, super-heated by the granite and rich in dissolved minerals, circulated through cracks, altering the hard granite on either side to a much softer and more crumbly rock. Once the granite was exposed, these weaker, hydrothermally altered zones eroded more rapidly, forming valleys which were excavated still further by the Ice Age glaciers. Gleann Einich, the Lairig Ghru, Glen Avon, Glen Derry and the Lairig an Laoigh are all thought to follow the lines of hydrothermal alteration zones later exploited by erosion.

On a smaller scale, the horizontal and vertical cracks in the granite, known as 'joints', and familiar to rock climbers, are also products of the geology and geological history of the rock. The vertical joints were formed during cooling of the granite, the rock shrinking and cracking as it cooled. The horizontal joints were formed much later, and result from the release of pressure which occurred as the rock above the granite was eroded away. As the weight of rock above was removed, the granite expanded upwards and cracks formed parallel to the upward-expanding surface. The 'horizontal' joints are, in fact, not all truly horizontal but follow the gentle contours of the pre-glacial land surface and are truncated by later glacial erosion.

Before the Ice Age

By Cretaceous times, much of the huge mountain chain uplifted during the continental collisions that closed the Iapetus Ocean had been reduced to a landscape of low relief near to sea level.

During the early Palaeogene, beginning about 65 million years ago, rapid uplift of different parts of the Highlands by as much as one kilometre accompanied the opening of the North Atlantic Ocean to the west of the Hebrides. Later more modest uplift may have continued on into the early part of the Ice Age, eventually elevating the Cairngorms to their present altitude. The early Palaeogene was also a time of strong global warming and the climate in Scotland was similar to the hot and humid subtropics in parts of western Africa today. This provided ideal conditions for deep chemical weathering of the bedrock.

Less intensive weathering continued during the cooler climates in the Neogene and into the early part of the Ice Age. Together, the uplift, weathering and erosion of the initially low relief landscape produced a series of plateau surfaces, rising in steps inland from the coast of north east Scotland and culminating in the Cairngorms.

Remnants of this pre-glacial landscape include the broad, sweeping plateaux which epitomise the Cairngorms, with their dome-shaped summits, sometimes capped with tors, and shallow river valleys such as Coire Domhain and the valley of the Feith Buidhe between Cairn Gorm and Ben Macdui. Locally, also, on the plateaux, pockets of weathered bedrock have survived glacial erosion, as in Coire Raibert between Cairn Gorm and Cairn Lochan, where the decomposed granite can be crumbled in the hand.

The Cairngorm tors are the finest in Scotland. These distinctive upstanding masses of granite are best seen on Beinn Mheadhoin, Ben Avon and Bynack More, where the Barns of Bynack reach a height of 20 metres. According to the classic explanation, the tors formed during pre-glacial times. Under the warm humid climate, areas of densely jointed granite were decomposed by deep sub-surface chemical weathering, leaving intervening areas of less densely jointed rock more intact. The weaker rotted rock was later eroded away, with the sound rock remaining as tors. However, it now seems unlikely that the present tors are truly pre-glacial remnants but are nevertheless of some antiquity, with the oldest possibly dating from around a million years ago. Probably shallower weathering of the granite during milder phases early in the Ice Age, followed by repeated removal of the weathered material, played a significant part in their development. The survival of the tors suggests that the summit plateaux on which they occur were not heavily eroded by the glaciers.

The shallow plateau valley of Coire Domhain, east of Cairn Lochan, contrasts with the glacially eroded cliffs of the Shelter Stone Crag at the head of Loch Avon

15

Into the Ice Age

The Antarctic Peninsula today gives an impression of how the Cairngorms may have looked during the coldest parts of the Ice Age

Over the last 50 million years, following the warming in the early Palaeogene, the global climate has progressively cooled. This cooling intensified about three million years ago, culminating in the start of the Ice Age around 2.6 million years ago when continental ice sheets expanded on the land masses around the North Atlantic.

Accumulation of permanent snow cover in the mountains led to the formation of glaciers on the plateaux and in the corries and glens of the Cairngorms and elsewhere in the Highlands. The glaciers moved slowly out over the lowlands to create huge ice sheets. This process is called glaciation. We now know from studies of the layering in the Greenland and Antarctic ice sheets, and from sediment cores drilled from the floors of the world's oceans, that the Ice Age consisted of many glacial periods separated by short warmer intervals, called interglacials.

During the interglacial periods, temperatures rose to levels similar to, or slightly higher than, those of today. Undoubtedly the mountains of Scotland were occupied by glaciers many times during the colder episodes and sometimes these glaciers expanded to cover the entire landscape.

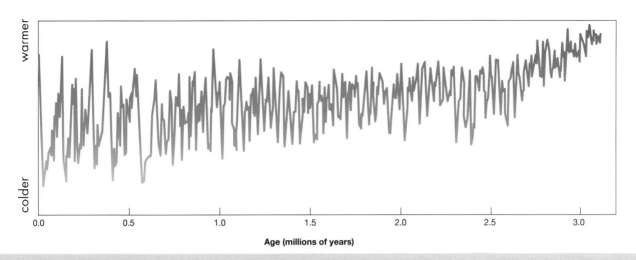

Age (millions of years)

Climate change during the Ice Age interpreted from a sediment core from the floor of the North Atlantic. The climate has switched regularly between colder and warmer conditions

However, because successive glaciations removed the traces of earlier events, the geological record on land is far from complete. The last major glaciation reached its peak around 22,000 years ago or slightly earlier when a large ice sheet covered all of Scotland and extended southwards to the Midlands of England. Glaciers extended out from the Cairngorms and merged with external ice flowing around the massif from the south west along Strathspey and into Glenmore and upper Deeside. By about 14,000 years ago most, if not all, of the ice had melted, producing dramatic landscape changes. The climatic warming was short-lived, however, and intensely cold conditions returned briefly after about 12,900 years ago and small glaciers again formed in the Cairngorms. Finally, around 11,500 years ago, rapid climatic warming initiated the present interglacial period, known as the Holocene.

During less cold periods, smaller glaciers existed in the Cairngorms, like this modern corrie glacier in northern Sweden

Glacial Sculptures

The Cairngorms are a classic landscape of selective glacial erosion. The extensive plateau surfaces contrast markedly with the deeply eroded glacial troughs, such as Glen Avon

While the broad outlines of the Cairngorms today reflect the character of the geology and the pre-glacial relief, the Ice Age has also left a distinctive imprint on the landscape. This is evident in a variety of landforms produced by glaciers.

We now think that the Cairngorms were completely covered by ice during the last major glaciation between about 30,000 and 15,000 years ago. This and earlier glaciations have produced a series of deeply eroded glens and troughs, with steep cliffs, cutting through the gently rolling pre-glacial plateau surfaces; Glen Avon and Gleann Einich are particularly striking examples. In places, the powerful glaciers also carved through the pre-existing watersheds to form a series of spectacular glacial breaches – the Lairig Ghru, Pass of Ryvoan, upper Glen Feshie, Inchrory and the Lairig an Laoigh. These breaches also diverted the pre-glacial headwaters of the River Feshie and the River Don. Other features of glacial erosion include truncated spurs, such as the Devil's Point, where the pre-glacial glens were straightened by the glaciers.

The contrast between these deep, ice-eroded glens and the rounded summits, tors and weathered bedrock on the gently rolling plateau surfaces is quite remarkable and is explained by the characteristics of the glaciers. Those on the plateaux were generally thin, frozen to the underlying bedrock and slow-moving. They were therefore incapable of much erosion, resulting in limited modification of the plateau surfaces and the unexpected survival of tors in a glaciated landscape.

In contrast, the glaciers in the glens were thicker, faster flowing and sliding over the underlying bedrock. Their power and speed made them very effective in eroding the granite, crushing and removing fresh as well as weathered bedrock. The glaciated landscapes of Baffin Island and East Greenland provide close present-day analogies.

The glaciers also shaped the bedrock in the larger glens around the Cairngorms. Asymmetric bedrock forms, known as 'roches moutonnées' are smoothed and streamlined on their up-glacier sides but end much more steeply and abruptly on their down-glacier sides, where weakened rock has been plucked away by the ice. They vary in size, ranging from rock knolls a metre high to hills a few hundred metres high. Good examples of small roches moutonnées occur near Dulnain Bridge where there is an interpretation site. Larger features include the hill of Ord Ban above Loch an Eilein and Farleitter Crag near Kincraig.

Corries are one of the most striking features of glacial erosion in the Cairngorms. These steep mountain basins were formed by small glaciers during episodes of less extensive glaciation. Their locations on north and east-facing slopes reflect the influence of wind-blown snow and shading on the development of mountain glaciers.

Meltwater channels on the northern flanks of the Cairngorms north east of Creag a' Chalamain. These channels formed at or beneath the margin of the last ice sheet as it melted in Glenmore

Rapid warming of the climate brought a dramatic end to the last ice sheet that covered much of Scotland. But the warming was not constant, and the sediments left by the glaciers record stillstands and re-advances of the ice, marked by fragments of moraines and other landforms at the glacier margins. As the glaciers melted, enormous volumes of water were released. Powerful meltwater rivers flowing under the ice cut into the underlying bedrock to form impressive channels that run across the hillsides. You can scramble through the rough bouldery gaps below Creag a' Chalamain and Airgiod-meall in upper Glenmore, and walk the meltwater channel of Clais Fhearnaig, which links Glen Lui and Glen Quoich in upper Deeside.

The meltwaters also carried vast amounts of sands and gravels. Some of these sediments were deposited under the ice on the flatter valley floors as eskers (sinuous ridges), and as kame terraces perched on hillsides alongside the glacier margins. Walking up the northern slopes of Bynack More gives a good view to the north end of Strath Nethy where you can see an impressive flight of kame terraces on the hillside. These formed when meltwater from the Glenmore glacier spilled over the ridge into Strath Nethy, cutting several deep meltwater channels and piling up sediment next to a smaller lower lying glacier. As this smaller glacier retreated, new lower terraces were formed.

Huge braided rivers discharged tonnes of sediment and water from the fronts of the glaciers and filled the valley floors with thick spreads of sand and gravel. These were later dissected by the rivers but their remnants survive as terrace fragments high above the modern floodplains in Strathspey, Glen Feshie, Deeside and Strathdon. Some terraces are pock-marked by kettle holes, which formed where blocks of ice became detached and abandoned in the river gravels, and later melted out. There is a trail guide for the walk around Uath Lochans at Farleitter Crag, near Kincraig, which describes the evolution of one large kettle hole.

Towards the end of deglaciation, some 15,000 years ago, glaciers were confined to the remote areas of the high plateaux, the headwaters of the River Dee and Geldie Burn, Gleann Einich, the Lairig Ghru and upper Glen Avon. In these places bouldery ridges and deposits mark many minor fluctuations in the ice margins, all cut through by meltwater channels. This would have been a desolate, dusty, grey landscape.

Thick accumulations of sand and gravel, seen here in a river bank section in Glen Feshie, were deposited by the melting glaciers

Meltwater channel on the northern flanks of the Cairngorms near Creag a' Chalamain

Ice-dammed Lakes

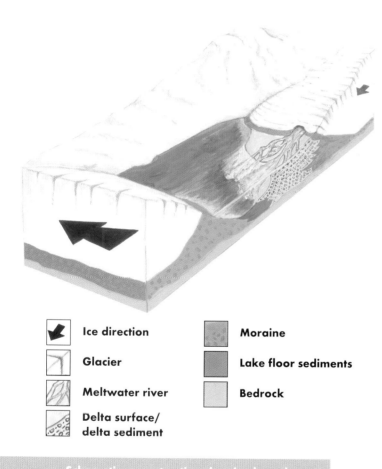

Key:

Ice direction	Moraine
Glacier	Lake floor sediments
Meltwater river	Bedrock
Delta surface/ delta sediment	

Schematic reconstruction showing how the terraces in Gleann Einich formed in a lake dammed between two glaciers

At the peak of the last glaciation, the huge glaciers that flowed down Strathspey, Deeside and Strathdon were joined by smaller, less powerful local glaciers flowing out of the Cairngorm Mountains. As all the glaciers became smaller, the Cairngorms glaciers became separated from the bigger valley glaciers, which blocked the exits for their meltwater rivers and so formed ice-dammed lakes. Evidence of these temporary lakes is found in Glen Quoich and Glen Derry in the south, and at the northern end of the Lairig Ghru and Gleann Einich in the north. However, the ice-dammed lakes did not all form at the same time. We know now from the sediments and the heights of delta terraces that the Lairig Ghru glacial lake formed slightly earlier than the Gleann Einich lake. Both were blocked by the Glenmore glacier, which also built thick accumulations of moraine along its margins.

Meltwater discharging from the local Cairngorms glaciers during the dramatic but short summer melt period brought large amounts of fine clays, silts, sands and cobbles into the lakes. The turbid lake waters would have been an opaque turquoise or grey colour, like modern glacial lakes. During the winter, ice formed on the lake surfaces and meltwater discharge into the lakes slowed down or stopped. During these quieter conditions, fine suspended sediment gradually rained down to the bottom of the lakes, forming thin layers of clay and silt on top of the coarser sediment deposited during the summer. At one site of a former lake in Gleann Einich, over 100 of these paired layers, known as varves, have been counted. This suggests that the lake could have survived for over 100 years.

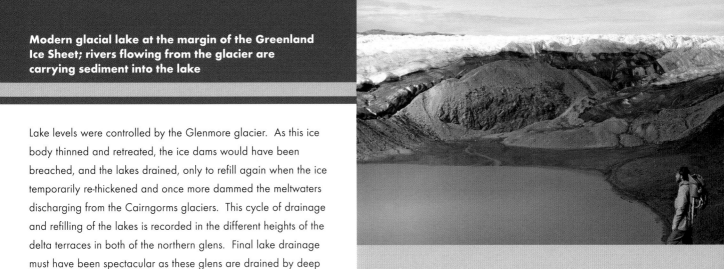

Modern glacial lake at the margin of the Greenland Ice Sheet; rivers flowing from the glacier are carrying sediment into the lake

Lake levels were controlled by the Glenmore glacier. As this ice body thinned and retreated, the ice dams would have been breached, and the lakes drained, only to refill again when the ice temporarily re-thickened and once more dammed the meltwaters discharging from the Cairngorms glaciers. This cycle of drainage and refilling of the lakes is recorded in the different heights of the delta terraces in both of the northern glens. Final lake drainage must have been spectacular as these glens are drained by deep gorges cut through the Glenmore glacier deposits.

These flat-topped delta terraces in Gleann Einich formed in an ice-dammed lake

Final Return of the Glaciers

Boulder moraines formed by a small corrie glacier in Coire an t-Sneachda, on the northern flanks of Cairn Lochan

The end of the last ice sheet glaciation, around 15,000 years ago, was marked by rapid environmental changes as the climate warmed rapidly from arctic to temperate conditions. These changes are recorded both in the surface landforms and in the sediments and pollen grains trapped in the deposits in lochs and bogs. As summer temperatures rose to levels similar to those of today, pioneer grass and sedge-dominated plant communities initially developed on the bare mineral surfaces. These were succeeded by shrub tundra with crowberry, birch and willow.

However, around 13,000 years ago the climate again deteriorated and glaciers returned briefly once more to the Cairngorms. This glacial event is called the Loch Lomond Readvance, after the area where it was first studied and named. The climate is thought to have been very cold and dry, judged partly on the abundance of mugwort *Artemisia* pollen, with average summer temperatures about 5 - 6°C below those of the present. Small glaciers returned to some of the high north and east facing corries in the Cairngorms and possibly the heads of a few of the major glens. The extent of these glaciers is indicated by boulder moraines and spreads of boulders in some of the corries, for example Coire an t-Sneachda just west of the Cairn Gorm ski area. Some moraines in the uppermost part of Glen Geusachan and at the head of Loch Avon may also have been formed at this time.

During this cold episode, the vegetation was sparse. Soil erosion increased on the bare mountain slopes and the frozen ground conditions encouraged the formation of solifluction features, rockfalls and scree slopes. The presence of plants such as dwarf willow *Salix herbacea* and the moss *Polytrichum norvegicum* indicates the development of extensive snow beds, while the remains of other plant communities suggest that the soil was disturbed or the ground was bare. This cold episode lasted for only about 1,500 years, and the climate again warmed rapidly 11,500 years ago when the last small glaciers finally melted.

We can thus tell from the record of the landforms and plant remains that the Ice Age did not end with a smooth transition from glacial to interglacial conditions, but with a series of rapid climate changes.

Landforms in a Cold Climate

Blockfield on Derry Cairngorm

Frost action in the soil and rocks has formed a range of distinctive 'periglacial' landforms (formed under cold, non-glacial conditions) on the upper slopes and plateaux of the Cairngorms during and since deglaciation. On several of the summits, including Ben Macdui and Derry Cairngorm, there are large areas of broken rock or blockfields which formed as frost and ground-ice shattered the granite bedrock. On steeper slopes, such as those of Lurcher's Gully in upper Glenmore, the loosened boulders and soil have been moved downslope by solifluction (soil creep/flow) processes, forming distinctive sheets, terraces and lobes of debris and boulders. The larger boulder lobes were last active at the time of the Loch Lomond Readvance, but some smaller solifluction features have been moving during the last few thousand years.

Today, the Cairngorms are well known for extremes of mountain weather, such as very high wind speeds, freezing temperatures, sudden summer storms and prolonged wet or snowy storms in winter. Under these conditions, periglacial processes are still active on the higher slopes and plateaux.

Wind-patterned vegetation on Carn Crom, between Glen Derry and the Luibeg Burn

Although smaller than the landforms formed at the end of the last glaciation, modern active features include sorted stone stripes, sorted circles, turf-banked terraces, 'ploughing' boulders, wind-eroded (deflation) surfaces and wind-patterned vegetation.

Prolonged wet weather and intense rainstorms can rapidly saturate and overload the soil, causing debris flows on steep slopes. These have left many slopes scarred by gullies and debris chutes, particularly in the Lairig Ghru. The eroded material often forms cone-like piles of debris at the base of the slopes, some of which in upper Glen Feshie have been forming for thousands of years.

Snow drifting off the plateaux during winter and spring forms cornices above steep slopes, which increases the likelihood of avalanches. In Coire an Lochain the granite slabs on the lower headwall are the source of dangerous slab avalanches. In the Lairig Ghru and on the western side of Derry Cairngorm, avalanches have swept the screes into tongues of boulders, characteristically littered with sharp angular fragments of rock.

Late-lying snowpatches occur in sheltered locations. The most persistent example is in An Garbh Choire above the Lairig Ghru between Braeriach and Cairn Toul, and there has been some speculation that it might have formed a small glacier during the so-called 'Little Ice Age' in the 17th and 18th centuries. Indeed travellers at that time reported large permanent snowbeds in the Cairngorms.

Boulder lobes in Lurcher's Gully, upper Glenmore

After the Ice

After the last glaciers melted, the sand, gravel and mud dumped by the retreating ice were easily eroded. However, as plants recolonised the bare, unstable soils, they slowed the pace of landscape change, through the binding effect of their root systems, which helped the soil and underlying sediments to withstand the erosive effects of running water and desiccating winds. Pioneer grasses and sedges were followed by dwarf shrubs and birch scrub. By about 9,700 years ago, birch and hazel woodlands were established in the glens, with grass and heath on the higher slopes. Pine trees increased around 8,800 years ago and came to dominate the natural forest on the more acid, well-drained soils up to a maximum tree-line altitude of about 800 metres.

Debris flows have extensively modified the slopes of the Lairig Ghru

Studies of the fossil remains of these trees preserved in peat bogs tell us that there were four periods of significantly wetter weather between about 8000 and 3,500 years ago. It seems likely that waterlogged soil and the resultant expansion of blanket bog, particularly after about 6,000 years ago, caused the pine forests to die back from wetter sites, which is why it is common to find old pine stumps in the peat bogs.

The glaciers left behind many unstable rock slopes. Screes accumulated, blanketing steep slopes below cliffs. In turn, these have been reworked by snow avalanches and debris flows. Where the screes have buried their crags, rockfalls have ceased, and vegetation taken over, such as in the Pass of Ryvoan. However, a small number of big crags collapsed in massive rockfall events, leaving large boulders jumbled into piles on the floors of Strath Nethy, the Lairig Ghru and at Loch Etchachan. The exact ages of these rockfall deposits are unknown, but elsewhere in Scotland similar large rockfalls occurred thousands of years after the glaciers had gone.

In the glens, kettle holes and hollows have been colonised by water-loving plants. Over thousands of years this has led to thick accumulations of peat in many hollows, reaching over three metres depth at Uath Lochans near Kincraig. As these peats accumulated, they captured pollen from the surrounding plants, forming an archive of the changes in the natural vegetation cover and land use.

Wandering Rivers

The steepness of the river channel, the amount of water and the availability of sediment that can be carried by the water are all factors which control river behaviour. The balance of these factors determines how much rivers can build up or erode their beds, and the type of channel patterns and channel changes that they display. Modern rivers in Scotland are less powerful than their glacial predecessors, and can only erode and transport gravel and larger rocks during floods. However, the thin soils, steep slopes and largely impervious bedrock of the Cairngorms allow rainfall to be channelled rapidly into the rivers and burns. The scale and frequency of flooding are therefore important in determining how river landscapes evolve.

Studies of the flood histories of a number of Cairngorms rivers reveal two characteristic types of flood: sudden, spectacular floods associated with summer storms, more typical of continental countries, and more prolonged periods of flooding following long periods of wet weather combined with snowmelt.

Braided rivers are now rare in Britain, with the best examples found in Glen Feshie where the glen is steep, flash floods are frequent and the river is actively undercutting slope and terrace deposits. Flash floods are also important in the evolution of steep boulder-strewn mountain torrents like the Allt Mór and the Luibeg Burn, and in active alluvial fan formation. Most Cairngorms rivers have a wandering, divided channel pattern, that can change abruptly during floods as gravel bars and scoured pools migrate, and new river banks become undercut.

People have used these rivers for transport and, up to the Second World War, for flaoting logged timber downstream. Rivers have also been altered by straightening reaches and strengthening river banks as part of flood control measures. Many of these old works, however, have fallen into disrepair, and been obliterated by the continued shifting of the river channels.

The meandering Derry Burn, Glen Derry

The Allt Mór mountain torrent, where a flash flood in 1978 swept away the bridge on the access road to the ski slopes on Cairn Gorm

31

Soils

Soils are an important component of the landscape and natural heritage of the Cairngorms. The soil types in the Cairngorms area reflect the geological legacy seen in the underlying glacial deposits and bedrock, and the local climate patterns, especially the changes in rainfall and temperature with elevation. These factors lead to the formation of three principal soil groups in the Cairngorms: podzols, montane soils and peat. Other less frequently occuring soils include gley soils, brown earths and alluvial soils.

Throughout much of the area, where rainfall is high, soils are very prone to 'leaching', which is the loss of material to water seeping through the soil. This leads to the development of the soil type known as a 'podzol'. At higher altitudes, where temperatures are lower, thin and fragile 'montane soils' are extensive. Peat formation is widespread where conditions are wetter, the ground is poorly drained, and dead vegetation rots very slowly. Where the ground does not drain properly, soils are often waterlogged most of the year and become 'gley' soils. Good agricultural soils (as represented by brown earth soils) are rare in the Cairngorms and mostly located in the straths.

The Cairngorms area is exceptional for an unusually large extent of rare, undisturbed soils. This is because the activities of people have had a less significant impact than in other parts of Scotland. However, soil is a non-renewable resource, continuously changing under human pressures and natural processes. Soil erosion may be accelerated by footpath expansion, trampling by people and deer, and also overgrazing. Peat erosion is thought to contribute to greenhouse gas emissions and climate change.

The land in the Cairngorms area is now mostly used for rough grazing, moorland, semi-natural woodland or commercial forestry, although there was formerly greater use of land for crops. The wide variety of semi-natural soils in the Cairngorms supports nationally and internationally important habitats and species and contributes to the remarkable landscape and natural beauty of the area.

Podzols

Podzols are prone to leaching and develop over acid parent material. They are easily recognised by the sharp contrast in colour between different layers in the soil.

Humus-iron podzol

The degree of leaching and accumulation of peat at the surface vary along an altitudinal gradient from foothills to summits, leading to the formation of different sub-types of podzol.

altitude ↑

Alpine podzol on summits

Peaty podzol and peaty gley podzol

Humus-iron podzol

Iron podzol and humus-iron podzol on lower foothills

Montane soils
and peat

Podzols and peat

Gley soils, peat and
occasional brown forest soils

Soil patterns in the landscape: Loch an Eilein and the Northern Cairngorms

Montane soils

Alpine (>700-850m) and subalpine soils are very slow to form, and are often shallow with a very thin organic layer over barely altered parent material.

Surface vegetation and roots anchor and protect the soil on steep slopes.

If the vegetation is damaged, soils become especially sensitive to trampling, physical damage and erosion.

Peat

Peat forms in poorly drained areas (such as sheltered parts of exposed summits or local basins) or areas of high rainfall.

A very slow rate of vegetation decomposition leads to the accumulation of dead plant material at the surface.

Peat is a rich natural store of carbon in the UK and is important in climate change scenarios.

Peat

Gley soils

'Gleying' occurs when downward movement of water is restricted, for example over compacted soil or impermeable geology.

Gley soils have waterlogged horizons and mottle-coloured patches where the soil has become deprived of oxygen. They are also often associated with foul, rotten smells.

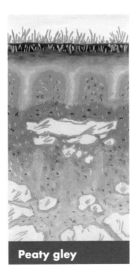

Peaty gley

Subalpine podzol

People in the Cairngorms

The earliest indication of people in the Cairngorm Mountains is a 5,000-year old flint arrowhead, found high on the plateau. Pollen records show that forest clearance took place at Loch Garten around 4,000 years ago, and some 2,000 years earlier at Howe of Cromar, near Aboyne, where there are remains of prehistoric settlement. Cultivation in the glens is also recorded by increased erosion due to forest clearance, and the appearance of pollen from cereals and agriculturally introduced weeds. Human activity has left an increasing mark on the landscape particularly during the last few thousand years.

More recent generations have also made significant use of the natural resources of the region. By the early 17th century, and continuing through the 19th century, people were mining local limestone. The limestone was heated in kilns to produce 'quick lime' which was spread on the land as a natural fertiliser. Limestone is abundant in a broad band of Dalradian rock running through Tomintoul and Braemar. However, limekilns are found across a much wider area, for example at Loch an Eilein where a smaller limestone outcrop was mined, and at Pass of Ryvoan where limestone may have been transported from some distance away.

Enjoying the Cairngorms landscape by bicycle

Old mine buildings at Lecht Mine near Well of the Lecht

Smoky quartz or 'Cairngorm crystal' 80g 6x3.5 centimetres

The main fuel for the lime kilns was local peat which was also used for domestic cooking and heating. Peat mosses near settlements and kilns were often worked out completely, but the hill peat continued to be an apparently unlimited resource. Today, traditional peat cutting for domestic use is no longer widely practised, although there is a commercial peat extraction operation near Tomintoul.

Also near Tomintoul, iron was mined in the hills near Well of the Lecht in the 1730s, with the ore being taken by pack horse to Nethy Bridge for smelting. In the 1840s, Lecht Mine was reopened to produce manganese ore, which was sent to Newcastle for use in the bleach trade. Alongside the more common iron and manganese oxide minerals at Lecht Mine, there are unusual zinc-manganese and lithium-manganese minerals. These minerals are part of an ancient soil, formed from iron- and manganese-rich Dalradian rock, which was deeply weathered in the tropical climate before the Ice Age.

Another mineral, famously found it the Cairngorms and collected extensively by the Victorians, is semi-precious smoky quartz, known as 'Cairngorm Crystal' or 'Cairngorm Stone'. Large and decorative samples of Cairngorm Crystal have been found mostly in late granite veins cutting through the Cairngorm granite.

Today, many rare mineral sites in the Cairngorms are protected from exploitation, and the widest use of the geodiversity of the region is by visitors who come to see and enjoy the very special landscape.

A Special Place

Ski-touring in the Cairngorms

The Cairngorm Mountains are a very distinctive area of Scotland, redolent of the Arctic in terms of their landforms, weather and plants. These mountains are on a grand scale, wide expansive landscapes, with large tracts of the higher land having survived from before the Ice Age, next to deeply eroded glens and corries cut by the glaciers. Together they form an outstanding example of a landscape of selective glacial erosion, which is a valuable scientific, educational and environmental resource.

This is also a dynamic, living landscape, underlain by granite and metamorphosed rocks, sculpted into distinct landforms by past and present processes and with a range of soils, plants and animals. It is a prime illustration of geodiversity supporting biodiversity. The patterns of vegetation communities reflect the interplay between soils, landforms, exposure, snow-lie and climate. The high-level plateaux are rich areas of montane vegetation, with lichen-rich montane heath and the specialised vegetation of snow-beds and springs, which includes many rare species.

The glacial deposits of the lower ground form well-drained ridges of sands and gravels interspersed with wet hollows. They support a mosaic of habitats for birds, insects and plants, and the largest areas of native Caledonian pinewoods in Scotland. The diversity of the Cairngorms landscape also provides many outdoor recreation opportunities. Summer and winter walking, ski-touring, and downhill skiing all take advantage of the unique glaciated granite landscape, while rock and ice climbing depend on jointed rock and steep, glacier-cut cliffs.

The Cairngorm Mountains have an exceptional value among Scotland's great wealth of natural heritage. International and national recognition of their conservation interests has resulted in many layers of official conservation designations, all intended to protect this impressive and special landscape and the plants and animals that thrive within it. Much of the high plateaux and the mountain glens are incorporated within Site of Special Interest (SSSI) and National Nature Reserve (NNR) designations. At an international level, a large part of the area is notified under the European Habitats Directive and the European Birds Directive. The Cairngorms area was also declared Scotland's second National Park in 2003. The international significance of the geodiversity of the Cairngorm Mountains is recognised by their inclusion in the UK Tentative List of World Heritage sites for their exceptional physical features. While this is impressive recognition, the most important means of conserving the natural heritage of the Cairngorms lies with all of us, in how we care for and use these majestic and ancient mountains.

The rocks, landforms and soils of the Cairngorms support a range of habitats for birds, plants and animals such as dotterel (left) and mossy saxifrage (right)

Exploring Geodiversity in the Cairngorms

To help you explore the geodiversity of the Cairngorms, the following set of trail guides provides an illustrated introduction to the landscape history of Glenmore and an area near Kincraig:

Trails Through Time in the Cairngorms National Park: Farleitter Crag Trail, Allt Mór Trail, Ryvoan Trail. Scottish Natural Heritage ISBN 1 85397 402 1 Price £2.50

Contact: Scottish Natural Heritage, Publications Unit, Battleby, Redgorton, Perth, PH1 3EW. Tel. 01738 458530, fax. 01738 458611 or email pubs@snh.gov.uk

The following publication gives a short account of the Cairngorm gemstones:
Cairngorm Stones. The Natural and Cultural History of Cairngorm Gemstones. Basil M.S. Dunlop. Grantown Museum and Heritage Trust. Price £4.00.

Contact: Grantown Museum, Burnfield Avenue, Grantown-on-Spey, Morayshire, PH26 3HH.
E-mail gosmuseum@btconnect.com.
www.grantownmuseum.co.uk

Further information on geodiversity and the Cairngorms can also be accessed via the internet:

www.fettes.com/Cairngorms/
Contains a wealth of information on the physical landscape of the Cairngorm Mountains, including sections on geology, tors and glacial landforms.

www.scottishgeology.com
This is the place to find out more about the geology of Scotland and see how the Cairngorms fit into the wider picture.

www.snh.org.uk
The website of Scottish Natural Heritage. It includes an introduction to Scotland's geodiversity and geodiversity conservation issues; information on sites protected by conservation designations; and educational material on Scotland's natural heritage.

www.cairngorms.co.uk
The website of the Cairngorms National Park Authority. It contains information on access and the latest news from the Cairngorms National Park

Location Map

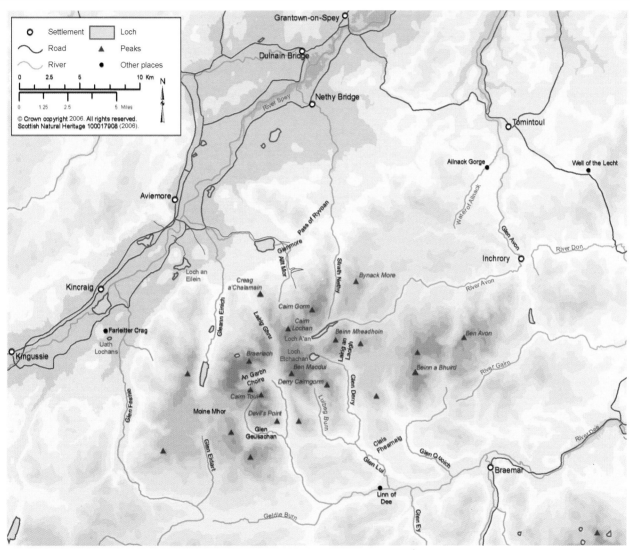

Legend:
- ○ Settlement
- ⬭ Loch
- 〰 Road
- ▲ Peaks
- 〰 River
- ● Other places

0   2.5   5   10 Km
0   1.25   2.5   5 Miles

N

© Crown copyright 2006. All rights reserved.
Scottish Natural Heritage 100017908 (2006).

Grantown-on-Spey

Dulnain Bridge

River Spey

Nethy Bridge

Tomintoul

Ailnack Gorge

Well of the Lecht

Water of Ailnack

Aviemore

Pass of Ryvoan

Glenmore

Allt Mor

Strath Nethy

Glen Avon

River Don

Inchrory

Loch an Eilein

Creag a'Chalamain

Bynack More

River Avon

Kincraig

Gleann Einich

Lairig Ghru

Cairn Gorm

Caim Lochan

Beinn Mheadhoin

Ben Avon

Farleitter Crag

Loch A'an

Lairig an Laoigh

Uath Lochans

Braeriach

Loch Etchachan

Beinn a Bhuird

River Gairn

Kingussie

An Garbh Choire

Ben Macdui

Derry Cairngorm

Glen Derry

Moine Mhor

Cairn Toul

Devil's Point

Lochan Uaine

Glen Feshie

Glen Geusachan

Glen Eidart

Luibeg Burn

Clais Fhearnaig

Glen Quoich

River Dee

Braemar

Glen Lui

Linn of Dee

Geldie Burn

Glen Ey

39

Scottish Natural Heritage
and the British Geological Survey

Scottish Natural Heritage is a government body. Its aim is to help people enjoy Scotland's natural heritage responsibly, understand it more fully and use it wisely so that it can be sustained for future generations.

Scottish Natural Heritage
Great Glen House
Leachkin Road
Inverness IV3 8NW

SCOTTISH
NATURAL
HERITAGE

The British Geological Survey maintains up-to-date knowledge of the geology of the UK and its continental shelf. It carries out surveys and geological research.
The Scottish Office of BGS is sited in Edinburgh. The office runs an advisory and information service, a geological library and a well-stocked geological bookshop.

British Geological Survey
Murchison House
West Mains Road
Edinburgh EH9 3LA

British
Geological Survey
NATURAL ENVIRONMENT RESEARCH COUNCIL

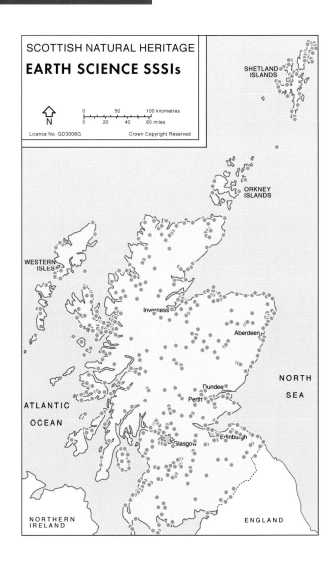

SCOTTISH NATURAL HERITAGE
EARTH SCIENCE SSSIs

N

| 0 | 50 | 100 kilometres |
| 0 | 20 | 40 | 60 miles |

Licence No. GD3006G   Crown Copyright Reserved

SHETLAND ISLANDS

ORKNEY ISLANDS

WESTERN ISLES

Inverness

Aberdeen

NORTH SEA

Dundee
Perth

ATLANTIC OCEAN

Edinburgh
Glasgow

NORTHERN IRELAND

ENGLAND

Remember the
Geological Code!

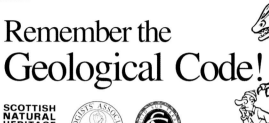

 SCOTTISH NATURAL HERITAGE

No need to hammer indiscriminately!
Never collect from walls or buildings.

Keep collecting to a minimum: remove fossils, rocks or minerals only when essential for serious study. And remember to refer good finds to local museums.

The leader of a field party should ensure that the spirit of the code is upheld.

Always seek permission before entering private land.

No one has the right to "dig out" any site. Try to leave the site as you found it!

Don't litter fields or roads with rock fragments, and avoid disturbing plants or wildlife.

Back fill excavations where necessary to avoid injury to people or animals.

Be considerate, and do not make things more difficult or hazardous for others coming after you.

Don't disfigure rock surfaces with brightly painted numbers, symbols or clusters of core-holes.

SAFETY FIRST!
✔ Wear protective goggles when hammering.
✔ Wear safety hats in quarries or below cliffs.
✔ Avoid loosening rocks on steep slopes.
✗ Do not get cut off by the tide.
✗ Do not enter old mine workings or cave systems.
✗ Do not interfere with machinery in quarries.

● Remember, you are one of several hundred geologists visiting this area every year — so your behaviour *does* matter.
● Please observe the code, so that others can also enjoy the great scenery, geology, and ecology here!

Published by Scottish Natural Heritage, 1996.

Also in the Landscape Fashioned by Geology series...

Arran and the Clyde Islands
David McAdam & Steve Robertson
ISBN 1 85397 287 8 pbk 24pp £3.00

East Lothian and the Borders
David McAdam & Phil Stone
ISBN 1 85397 242 8 pbk 26pp £3.00

Edinburgh and West Lothian
David McAdam
ISBN 1 85397 327 0 pbk 44pp £4.95

Fife and Tayside
Mike Browne, Alan McKirdy & David McAdam
ISBN 1 85397 110 3 pbk 36pp £3.95

Glasgow and Ayrshire
Colin MacFadyen and John Gordon
ISBN 1 85397 360 2 pbk 52 £4.95

Glen Roy
Douglas Peacock, John Gordon & Frank May
ISBN 1 85397 360 2 pbk 36pp £4.95

Loch Lomond to Stirling
Mike Browne & John Mendum
ISBN 1 85397 119 7 pbk 26pp £2.00

Mull and Iona
David Stephenson
ISBN 1 85397 423 4 pbk 44pp £4.95

Northwest Highlands
John Mendum, Jon Merritt & Alan McKirdy
ISBN 1 85397 139 1 pbk 52pp £6.95

Orkney and Shetland
Clive Auton, Terry Fletcher & David Gould
ISBN 1 85397 220 7 pbk 24pp £2.50

Rum and the Small Isles
Kathryn Goodenough & Tom Bradwell
ISBN 1 85397 370 2 pbk 48pp £5.95

Skye
David Stephenson & Jon Merritt
ISBN 1 85397 026 3 pbk 24pp £2.50

Scotland: the creation of its natural landscape
Alan McKirdy & Roger Crofts
ISBN 1 85397 004 2 pbk 64pp £7.50

Series Editor: Alan McKirdy (SNH)
Other books soon to be produced in the series include:
Ben Nevis and Glencoe, Western Isles

SNH Publication Order Form

Title	Price	Quantity
Arran & the Clyde Islands	£3.00	
Cairngorms	£4.95	
East Lothian & the Borders	£3.00	
Edinburgh & West Lothian	£4.95	
Fife & Tayside	£4.95	
Glasgow and Ayrshire	£4.95	
Glen Roy	£4.95	
Loch Lomond to Stirling	£2.00	
Mull and Iona	£4.95	
Northwest Highlands	£6.95	
Orkney & Shetland	£2.50	
Rum and the Small Isles	£5.95	
Skye	£3.95	
Scotland: the Creation of its natural landscape	£7.50	

Postage and packaging: free of charge within the UK. A standard charge of £2.95 will be applied to all orders from the EU. Elsewhere a standard charge of £5.50 will apply.

TOTAL _____

Please complete in **BLOCK CAPITALS**

Name _____

Address _____

_____

_____

Post Code

Method ☐ Mastercard ☐ Visa ☐ Switch ☐ Solo ☐ Cheque

Name of card holder _____

Card Number ☐☐☐☐ ☐☐☐☐ ☐☐☐☐ ☐☐☐☐

Valid from ☐☐ ☐☐

Expiry Date ☐☐ ☐☐

Issue no. ☐☐

Security Code ☐☐☐ (last 3 digits on reverse of card)

Send order and cheque made payable to Scottish Natural Heritage to: Scottish Natural Heritage, Design and Publications, Battleby, Redgorton, Perth PH1 3EW

pubs@snh.gov.uk

www.snh.org.uk